BJÖRN IRONSIDE

Copyright © 2021
University Press
All Rights Reserved

Table of Contents

Introduction
Chapter 1: Björn's Parents: Ragnar Lothbrok and Aslaug
Chapter 2: A Viking Childhood
Chapter 3: Viking Homes and Ships
Chapter 4: Björn's Siblings
Chapter 5: Aslaug's Seidr
Chapter 6: Norse Religion
Chapter 7: Viking Rituals
Chapter 8: Ragnar's Jealousy and Death
Chapter 9: The Quest for Riches, Land, and Glory
Chapter 10: To Conquer Rome
Chapter 11: A Plan Fulfilled?
Chapter 12: Bjorn Ironside, King of Sweden
Chapter 13: A Viking Funeral
Chapter 14: The Munso Dynasty
Chapter 15: Björn Ironside's Story Lives On
Conclusion

Introduction

Stories handed down in the oral tradition for centuries and later recorded by monks tell the tale of a powerful Viking who set out to conquer Rome. Bjorn Ironside became a legendary figure for the Norse people even as his European exploits were taking place.

Today, little is known of Bjorn Ironside. His story is marked by inconsistencies and exaggerations, leaving the modern world to rely on conjecture to fill in the details of his life. Yet, we do know that Bjorn Ironside existed and that he eventually became King of Sweden.

In addition to the historical record, archeological digs have uncovered a wealth of information about Vikings like Bjorn Ironside. We know details of the Vikings' everyday lives, their beliefs, how they made their tools and built their houses and ships, how they fought, and how they celebrated their victories. We also know the challenges they faced and can make educated guesses about why they ventured out to explore, trade, loot, and pillage on a global scale.

Bjorn Ironside was a member of a Viking family well-known for striking fear in the hearts of their enemies and victims alike. His father, Ragnar Lothbrok, is the subject of intense controversy among Norse scholars, some of whom believe he was not a single person but a composite character based on several historical figures. However, if Bjorn Ironside existed and was the son of Ragnar, it stands to reason that Ragnar Lothbrok was also a real person. With little to go on, the question may never be answered.

So, while Bjorn Ironside's story includes many historically accurate facts, the best way to experience the power of the legend is to start with what we know is real and then follow the myths and legends into the shifting sands of historical probability.

As you read this biography of Bjorn Ironside, we encourage you to let your mind wander from what certainly was to what might also have been true. Dive into the legend with an open mind and a willingness to accept the historical possibilities presented here. You may find that you gain a new admiration for Bjorn Ironside or his actions may repel you. Either way, you will see for yourself why history took note of this incredible Viking chieftain.

Once upon a time, there was a great Viking leader who set out to own the most powerful city in the world. That man was Bjorn Ironside, and his legend survives him by over 1000 years. What will you make of Bjorn?

Chapter 1

Björn's Parents: Ragnar Lothbrok and Aslaug

No biography of Bjorn Ironside would be complete without the story of Bjorn's parents. His father, Ragnar Lothbrok, was among the most celebrated men of ancient Scandinavia. And, although many modern sources claim that Lagertha the shieldmaiden was Bjorn's mother, the Medieval texts refer to Aslaug as his mother. These two had remarkable lives as well as a unique shared history.

Ragnar was the son of Danish King Sigurd Hring and Alfhild, the daughter of King Alf of Alfheimer. So, he was a noble before he began his career of raiding European cities. He won the heart and hand of Lagertha by killing a bear and a hound she had sent to attack him. The couple had a son they named Fridleif and two daughters whose names are unknown. Although Lagertha's father had given Lagertha to Ragnar to be his wife, Lagertha was never happy with the arrangement. The marriage soon unraveled as Ragnar grew tired of Lagertha's anger towards him.

Ragnar acquired his nickname Lothbrok while courting Thora Town-Hart, who would become his second wife. As the story goes, Ragnar fought a snake to win Thora's hand. To protect himself from the snake's venom, he wore a suit of fur dipped in tar. These clothes earned him the name "Shaggy Breeches," which translates to Lothbrok in the Old Norse language. Thora and Ragnar had two sons, Eirikr and Agnar. Both of these sons died in battle. Thora eventually grew ill and died, leaving Ragnar on his own once more.

When Ragnar met Aslaug, a Norse queen, he immediately wanted to have her as his wife. Aslaug was the daughter of Sigurd, a mighty Viking warrior, and Brynilde the shield-maiden. When Aslaug was young, her parents were killed, and she was raised by a foster father. After the death of Aslaug's parents, the foster father was afraid for her safety. He put ragged clothes on her to disguise her beauty, put her in a harp, and journeyed with her far from her home. He called her Kraka, meaning Crow, to emphasize this newly-gained ugliness.

However, while she was bathing one day, some men found her and saw how beautiful she actually was. When they arrived back at their ship, commanded by Ragnar Lothbrok, they told Ragnar about the beautiful woman they had found. He wanted to see her for himself and sent a message to her that she must come to him dressed and undressed, alone and not alone, and neither fasting nor eating. She arrived dressed in a net, biting an apple, and with a dog for a companion. Ragnar Lothbrok was so impressed with her cunning that he wanted her for his wife.

Aslaug was something of a seer and a sorceress. When she and Ragnar were married, she wanted him to wait three nights before consummating their marriage. She told him that if he didn't, his son would be born without bones. He would not wait, and his son Ivar the Boneless was born soon afterward. She told him later that he would have a son born with the image of a snake in his eye. Thus, Sigurd Snake-in-the Eye was born. Bjorn Ironside was also a son of Aslaug, along with another son named Hvitserk.

Ragnar had a competitive spirit that extended to his sons. Although he was proud of their exploits, he was jealous of their victories and reputations. As the boys grew up, their father continually tried to outdo them.

Aslaug continued to make prophecies. Before Ragnar set out on his final voyage, she told him his fleet was not ready to make an invasion. Aslaug made him an enchanted shirt that would protect him from snake venom. He ignored her warnings and was not only defeated but eventually tortured and killed. Although there were failures and losses, Ragnar more often came out ahead after raiding a country.

Chapter 2

A Viking Childhood

Bjorn, born in Denmark about 777 CE, must have had a childhood similar to that of other Viking children of his time, place, and social position. While little is known specifically about Bjorn Ironside before he set out on his voyages, archeological research along with small bits of information in the larger historical record has revealed something of what life was like for them.

Aslaug was known as a strong woman. She survived childbirth, which was something of a feat in itself. At that time and place, women died much younger than men, simply due to complications of childbirth. On average, men lived to be 40 years old, while women only tended to live until about 35.

Bjorn beat the odds, too, by surviving a childhood set in the cold and harsh environment where he grew up. Had he been born sick or disabled, he would have been taken far from the Viking settlement and left to die. Even those born healthy were vulnerable to diseases and illness for which there were no cures. About 20% of Viking children died before they were five years old. Another 20% never saw their 20th birthday.

Still, Viking children had much pleasure in life. Their parents' first demonstration of love for them was a little Thor's hammer charm, which was thought to ward off evil spirits after they were born.

As they grew into childhood, they were given toys like hummers, which were noisemakers made of pig bone and twisted cord, or whistles made of geese bones. Children also played with wooden dolls and toy boats. They played football, too, and a sport played with a stick and ball called Knattleikr. They loved board games, including Knefatfl. They explored the countryside around their homes and listened to songs and stories, both in their homes and in the community, when traveling musicians and storytellers visited.

Yet, childhood was not all fun and games. Children had to work to help their parents on the family farm. The lower the family's status, the more work the children had to do. Nearly all Viking families owned slaves, who did the hardest and dirtiest jobs. Often the slaves came from cities the Vikings raided. They brought these people home and put them to work. Since many tasks needed to be done to keep a Viking family going, there was still plenty of work for the children of the family to do.

As a child of a King, though, Bjorn Ironside would have escaped much of the drudgery of farm work. Instead, he would have been taught to learn the Norse runic alphabet along with other scholarly subjects like the law, religion, and history of his people. Viking children (as well as adults) did not read books but learned their lessons in the oral tradition.

Children also learned the skills they would need to survive in the world. Viking boys learned how to make tools, weapons, and homes. Since nearly every Viking man had a ship, the boys learned how to build them and how to navigate by the stars and landmarks. Boys, and to a lesser extent, girls, learned how to use swords. As the son of a famous Viking raider and King,

Bjorn would have received advanced instruction in the art of battle.

One unique skill that has been glossed over in contemporary accounts of the time is a martial art developed by the Norse people. It was called Glima, and young boys began studying it at the age of around 6 or 7. Glima helped Vikings react at lightning speed, even if they had no weapons available to them at the time they were attacked. This martial art was comparative to other martial arts practiced around the world. There were throws, kicks, locks, and defenses to them, just as there are in the more well-known Asian martial arts.

Glima, by the way, could be practiced as a sport in Glima wrestling, which was practiced as Brokartok, or trouser-grip, Hyggsooebba Glima, or backhold wrestling, and Lausatok or free-grip wrestling.

Unlike other forms of wrestling, Brokartok Glima wrestling demanded that the opponents stand erect. They had to step back and forth in a kind of a dance. They could not fall on an opponent or push him down. Finally, they had to look over their opponent's shoulder so they could fight by touch and not by sight.

Hryggspenna is much like current forms of wrestling. While the other types of Glima wrestling were all about technique, this form of Glima was more of a test of strength.

Lausatok Glima is still practiced today and is known as Viking wrestling. It features exceptionally aggressive moves. It can either be practiced for self-defense or for competition. It is called 'free-hold' wrestling because the opponents can use any type of hold.

Viking children like Bjorn Ironside often practiced Glima and became very proficient at it. It was not only a way to learn fighting skills, however. Another reason it was taught to young boys was that it gave them great self-confidence.

Children were expected to grow up quickly. By the age of 5, they began learning all those valuable life skills, and by 10, children were treated more like adults than children. Then, before they reached the age of 16, they were expected to go out on their own; usually they acquired farms, often after a time spent trading in the markets or raiding prosperous cities.

Since they had already learned how to build homes and ships and craft all the weapons they would need, young Vikings were well-equipped to set up homes and raiding expeditions by the time they left their parents' home.

Chapter 3

Viking Homes and Ships

Little remains of the homes the Vikings built in the time before, during and for several centuries after Bjorn Ironside's time, despite many archeological digs in lands where Vikings once lived. This is because they were made primarily of wood, which would decompose in the years between the Viking Age and the time archeological excavations began. Yet, we do know some facts about these wooden structures, both from the written records of the stories that were told and from certain clues in the excavation sites.

Most homes were long and narrow and were often called longhouses or halls. Several families would typically live together in one longhouse that might measure up to 100 feet long. Each longhouse had a fireplace and a smoke hole above it, located at or near the center of the house. A few small, glassless windows provided little ventilation. So, the longhouses would often become filled with fumes and smoke, despite the smoke holes.

The poorest Vikings did not have wooden homes. Theirs were made of a combination of sticks, dung, soil, and straw called wattle. These wattle homes were colder than longhouses and more vulnerable to damage in storms. With their thatched or wood-tiled roofs, the longhouses were about as cozy as life got in Medieval Scandinavia.

Eventually, nearly all Viking men left their homes to make trades in the global marketplace or raid other countries. They sought riches and lands. To accomplish their goals, they had to leave Scandinavia for other lands. Over time, they developed shipbuilding skills that people in other lands could only dream of.

All but slaves and extremely poor Vikings owned ships. This was necessary because so much of their homeland was covered in rivers, lakes, and fjords. To move around freely in Scandinavia, they needed boats. Later, their seaworthy vessels were used to travel beyond their homeland and into the wider world to trade or to conquer other lands.

Shipbuilding was a significant part of life in the Scandinavian countries, and they were considered the masters of the craft through the 1300s. They were extremely advanced in their ability to design boats fit for different needs within the Viking communities. They built ships for short-distance travel and fishing. These were called Karves. The boats they built for transporting cargo were called Knarrs. And then there were the longships, designed for extensive travel and raids.

All Viking ships were made with planks of oak or occasionally pine. The planks were riveted together, and the openings between them were filled in with animal hair. Viking longships had a unique structure that enabled them to travel great distances, even in the roughest seas. When a longship was put into the water, the planks would move individually and together, creating a flexible ship that would not break up in rough seas and high tides.

Vikings carved elaborate dragons or serpents on the fronts of their ships, often called Dragonships. Aside from artistic expression, the carvings were designed to ward off sea monsters and terrorize their enemies. These longships were used for long-distance exploration, trading, and raiding. When the people in other lands saw the longships with their terrifying carvings, they knew that a battle was at hand.

Longships were between 80 and 100 feet long, with enough room to hold up to 60 men. Knarrs were taller and wider than longships, but only about 50 feet long, allowing only enough space to carry a crew of about 30 men. Karves were narrower and shorter than Knarrs but a bit longer, at about 70 feet.

When the sea wind was still or when the wind was so strong that it threatened to push the ship over, the men rowed the ships with colossal oak oars. They sat on trunks that carried all their possessions. If conditions became dangerous, the sails were put down and used to cover the sailors in the ship below. However, when the winds were right, the large, square sails were up, billowing in the breeze and pushing them steadily onward.

Chapter 4

Björn's Siblings

When Bjorn came into the Viking world, he already had five step-siblings – Friedlief, Eirikr, Agnar, and at least two sisters. Eirikr and Agnar fought alongside Bjorn and his full brothers before Ragnar Lothbrok died. Friedlief is barely mentioned in the Tales of Ragnar's Sons, a Medieval text about the clan. The girls, too, have nearly been forgotten by history.

Bjorn's full brothers were very much a part of his life. As children, they likely played Knattleikr, competing with each other as they maneuvered their ball around a meadow with a stick. On winter days, they might have been found playing Knefatfl, placing their pieces on the board skillfully to best their opponent in the board game so familiar to Viking children. And, as Viking children in a society where shipbuilding and sailing were crucial, they probably made toy boats and sailed them down nearby rivers or fjords.

The boys would have been taught and cared for by their parents for the first few years of their lives. Then, one by one, as they reached the age of 5 years old, the sons of Aslaug and Ragnar would have been sent to study swordsmanship, the use of battle axes and spears, battle strategy and tactics, and how to make weapons and tools at the home of a relative or a respected member of the community.

They also would have learned all about ships, from building each type of ship to navigating them by the stars. As a prelude to building these all-important vessels, they would have practiced with wood and metal by making barrels and other items useful in their family home or worthy of sale at market. They may have been taken to market to learn how to trade the weapons and tools that they had made. However, because they were the sons of the great Ragnar Lothbrok, their instruction was likely geared more to battle than to trade.

Ivar the Boneless was the oldest of these brothers. The Medieval texts surrounding the Ragnar Lothbrok story tell how the other brothers, as well as his crew members, carried Ivar to his ships or out to the battlefront. When it came to being a mighty chieftain, though, Ivar was not disabled in any way. He commanded his men well and even joined in the battles. After many years of sailing, raiding, and looting, Ivar helped lead the Great Heathen Army in its quest to kill their father's murderer.

Bjorn, born around 777 CE, was likely the second son of Aslaug and Ragnar. Sigurd Snake-in-the-Eye, like his brothers, was known and feared as a ruthless warrior and Viking chieftain. Hvitserk was less famous than the other brothers, but he did go on at least one of Bjorn Ironside's journeys with him and fought alongside Bjorn and the other brothers. Halfdan Ragnarrson and Ubba, Bjorn's other two brothers, took part in the raids as well and were leaders of the Great Heathen Army. In fact, Halfdan took over command of that army after they conquered England and Ivar stayed behind to rule.

Ragnar Lothbrok was incredibly proud of his brave sons as they bested city after city in Scandinavia and then Europe. At the same time, he was jealous of his sons' success. He was continually trying to outdo them and show that he was the superior Viking chieftain. He wanted them to become rich and get land, but he also wanted them to be completely subordinate to him.

Bjorn and his brothers did acquire wealth. They gained land, as Viking men were expected to do. Their enemies panicked when they saw the ships of Ragnar's sons sailing near the coast. Just as Ragnar had done, they became well-known for their raids of Europe. In doing so, they both honored and frustrated their father, Ragnar.

Chapter 5

Aslaug's Seidr

Bjorn, son of Ragnar and Aslaug, was called "Ironside" because he was never injured in battle. As the story goes, Aslaug used seidr, a type of magic ritual, to make him invulnerable to steel and iron. In this sense, he was one of the mightiest Vikings who ever lived. He and his brothers shared many adventures during their lifetimes, but no matter what kind of scrape they found themselves involved in, Bjorn was never harmed by a sword, battle-ax or spear.

Aslaug, also known as Kraka, Randalin and the third wife of Ragnar Lothbrok, was known to be a seer and sorceress. Many of the stories surrounding the Lothbrok legend and the tale of Ragnar's sons recount incidents in which Aslaug used magic for good, to protect her loved ones in times of crisis.

When Ragnar decided to go to England to kill King Aella, Aslaug knew he would be in grave danger. She made him a beautiful shirt that would protect him as long as he wore it. Then, too, there was the seidr she performed to make Bjorn Ironside invulnerable.

Women were often the seers and sorcerers of the Old Norse world. A man could perform a seidr, but it was considered effeminate for him to do so. Typically, the sorceress performing the seidr would be assisted by several helpers. The women of the community also had to be present to lend their voices and their support. They all sang and chanted to get in touch with the gods and the spirits of their dead ancestors. They danced in a circle to invoke the spirits. The sorceress would speak an incantation and cast a spell. In this respect, the sorceress was much like the witches of folk tales from around the world.

Until recently, the stories of seidr were confined to fantastical tales and literary texts. However, recent archeological digs have found concrete evidence that seidrs were indeed conducted in Scandinavia at the time when Bjorn Ironside would have lived.

A seidr was a shamanic rite. The religious leader, or shaman, of the tribe would go on an inward, spiritual journey, taking along the other women of the community as she went through the experience. Often, magical items, such as Ragnar's shirt, were produced during the seidr. Other times, a community member or friend was blessed through the seidr; Bjorn's gift of invulnerability would have been a blessing bestowed in a seidr, for example.

However, seidrs were not always positive. Sometimes, they were done to put a curse on an enemy. They were also used to foresee the future. This would have been a very helpful power, because at that time, with no mass media, it was always uncertain what the Vikings would find when they reached their destination.

The seidr, although involved with the supernatural, was not the whole of the Vikings' religion. In fact, many aspects of the Nordic faith were based on ancient histories, old customs, and long-told stories. The Norse gods played a part, and the Vikings' religion had many unique customs to honor them and to plead for their help.

Chapter 6

Norse Religion

To say that the Vikings had a religion would be mischaracterizing their beliefs. In fact, there is no word for religion in the Old Norse language. The closest word is *sidr,* which can be loosely translated as "custom."

The Vikings were pagans in the sense that they were not Christians. They did not believe in their gods in the same way Christians do. They knew the stories of their gods, and they shared them often around the family hearth. Yet, they did not worship them as Christians do their God. The Norse gods were simply seen as a fact of Viking life. The gods had their faults, but on the whole, they were respected.

Although the people with the highest status in Viking culture usually conducted the religious ceremonies and rituals, there was no priesthood. Technically, anyone in the community was allowed to perform a rite. There were local customs, which were carried out mainly in family gatherings. Some evidence suggests that there were also societal customs, which were held in centralized locations and carried out by the king.

Local rituals took place in family homes or in large halls designed for this purpose. In addition to a hall for worship, there would also be a building where animals were sacrificed. The purpose of the Norse folk religion was to ensure the survival of the clan or the kingdom and regenerate society. Most customs followed a yearly cycle unless they were related to a life event that might happen at any time, such as a birth or a marriage.

There were two clans of Viking gods. The Aesir were war gods and sky gods. War gods like Odin and Thor were much revered by the Viking raiders. Vanir gods like Freyr and his twin, the goddess Freyja, protected the people and bestowed on them prosperity and fertility. The trickster god, Loki, was always up to some kind of mischief, making trouble between the Aesir and the Vanir. Still, the other gods typically won out.

The Norse custom was composed of many different stories, which varied greatly between the different locations. This is because the clans of Scandinavia were often isolated by the rugged terrain and the abundant waterways that cut through the region.

When Christianity was introduced to the Vikings, they called its practices the new customs or *nyr sidr*. Eventually, many Vikings accepted the new religion and began to practice it. The Christians thought of the ancient Norse customs as paganism – a name they gave religion that was not their own – and devil worship. Furthermore, the Christians wrote and spoke about the Norse religion as being evil, ignorant, and crude.

Unfortunately, no Norse texts written from the oral record exist that tell about the Vikings' folk religion from their perspective. Therefore, little is known about how the mythology of the region played into the Vikings' everyday lives. For the most part, the only records were in the Christian texts written long after the Norse customs were largely abandoned.

Chapter 7

Viking Rituals

Rituals were a part of family life and the life of the region. Major life events were celebrated with unique customs. The cycle of the seasons was also celebrated with its own unique rituals. There is no doubt that Bjorn Ironside took part in many rituals over the course of his life, and as a chieftain with high social status, probably conducted them. The Viking rituals can be separated into private rituals that were held among family members and public rituals any Viking could attend.

Private Rituals

Birth Rituals: When a child was born, the family performed rituals in a desire for the child to live and thrive. They prayed to the god Frigg and the goddess Freyja because these were the gods whom they believed could help the child most. They sang ritual songs and waited for nine days to pass. At the end of that time, if the child lived, the father would hold the infant on his knee, signifying that the child was now a part of the clan. If the child was abandoned or killed after that, it would be considered murder.

Marriage Rituals: Viking wedding ceremonies were elaborate and involved rituals that started with the proposal. It all started when a delegation headed by the groom visited the bride's family to make the proposal. After the groom and bride's family worked out details like the dowry, the groom's gift to the bride, and determined the inheritance rights, they held a large feast. The wedding feast was several days. Socially acceptable marriage feasts were at least three days long. The god Var was considered the witness to the bride and groom's vows. When the marriage ceremony was over, the guests led the bride and groom to bed for their first night together.

Public Rituals

Seidr: Although the Seid is mentioned in Medieval texts, little is known about the actual ceremony. It involved some type of magic practice and divination. Omens were interpreted with the goal of helping the Vikings or destroying their enemies. It is widely accepted that the Seid had some connection to the Norse gods, although the exact relationship is not known.

Blot: The blot was a blessing ritual. Over time, the word blot came to mean sacrifice. However, a story has been told that the blot could involve slapping a Norse woman in the face with flowers as a courtship blessing. And, of course, a large feast was an element of any blot.

Types of Worship

Vikings worshipped their ancestors. They believed their deceased ancestors could come back to bless them if rituals were carried out correctly or curse them if not. When someone died, a burial mound was built near the family home. Then, their grave was readily available to the family, who offered sacrifices to the dead or retrieve holy objects from the mound.

The Vikings also worshipped wights, which were deities that protected the land. Women likely conducted the ritual offerings to the land wights wherever they were found, whether near the family farm, in nearby groves or in waterfalls.

Sacrifices

Ritual sacrifices were a significant part of the Norse folk religion. Vikings like Bjorn Ironside would pour out wine or food to gain the favor of the gods. If there was going to be a marriage feast or a blot, animals were sacrificed to the gods and then eaten during the feast.

Some sources indicate the human sacrifice was a part of Norse paganism at the time of Bjorn Ironside and his brothers. Typically, a servant or Viking woman of low stature was sacrificed for a special blessing or to overcome a major challenge.

Chapter 8

Ragnar's Jealousy and Death

Ragnar Lothbrok wanted Bjorn Ironside and his other sons to succeed and continue the family tradition of bravery in battle. He wanted them to gain wealth and land. Yet, at the same time, he did not want their reputations to surpass his. In the end, Ragnar's jealousy would be his undoing. It would also lead to Bjorn and his brothers securing an even greater place in history than they might have otherwise. The story is told in the Medieval text *The Tale of Ragnar's Sons.*

Bjorn Ironside and his brothers lived in Sweden, but they were anxious to gain new lands. They started with nearby Zealand, Denmark, where they conquered Jutland and all the small islands in that area. Afterward, they took over Zealand and lived there for a time with Ivar the Boneless in charge of the country.

When Ragnar heard of his sons' success, he felt compelled to show he was greater than they were. He made Eysteinn Beli the Jarl of Sweden and told him to protect the country against his sons. With that, he set sail in quest of more riches and glory to show his superiority over Bjorn and his brothers.

However, Eric and Agnar had other ideas. They sent a clear message to Eysteinn that he must do as they and their brothers demanded. Furthermore, Eric wanted the hand of Eysteinn's daughter. Eysteinn was in no mood to deal with Ragnar's sons, so he consulted with the chieftains of Sweden to decide what to do. The chieftains said Eysteinn must fight back against Eric, Agnar and their brothers.

A great battle began, and soon Ragnar's sons were in peril. Agnar died in battle. Eric was taken prisoner, and Eystein offered to give him Uppsala to compensate him for the loss of Agnar. However, Eric refused the offer. Instead, he said he wanted to choose the day and the manner of his death. He asked that Eysteinn's men impale him on upright spears, placing him at a higher physical position than the dead. Eystein and his men complied with Eric's request. Now, two of Ragnar's sons were dead.

Bjorn Ironside was playing a game with his mother Aslaug and his brother Hvitserk to pass their time in Zealand. When they heard of Agnar and Eric's deaths, they were furious. Immediately, Bjorn and Hvitserk sailed to Sweden to take revenge. Aslaug, also known as Randalin, took a land route with the brothers' cavalry to join the great battle, where they killed Eysteinn.

Other fathers might have been proud of their sons for standing up for their brothers and protecting the family name. Ragnar, though, was jealous. Ragnar had expected that if his sons needed to fight, they would call on their father to lead them. In his anger, Ragnar sailed to England with two knarrs, the ships built for transporting cargo, not for war. His aim was to conquer England to prove his superiority over his sons.

Unfortunately, King Aella of Northumbria had other plans. His army defeated Ragnar Lothbrok and his men. Then, the king had Ragnar put in a pit of snakes to punish him for the offense.

Ragnar was fine as long as he wore the enchanted shirt Aslaug had made for and given him before his journey. However, when King Aella ordered the shirt removed, Ragnar was vulnerable to the snakes' bites. As he died, he sang a song, which was added to by his skald and recorded for history in the skaldic ode *The Death Song of Ragnar Lothbrok*, now called simply *Kraka's Song*.

To avenge Ragnar's death. Bjorn and his brothers attacked King Aella. They won the battle, and Aella gave Ivar the Boneless a large area of land. There, Ivar built the city of York. Ivar was popular among the English people, and when he asked his brothers to attack Aella again, many of the English chieftains and their people helped Ragnar's sons defeat Aella.

When Bjorn and his brothers had King Aella where they wanted him, they discussed how they should kill him. They decided that the only just punishment for the man who had tortured and killed their father was for them to carve the blood eagle on him.

The blood eagle is known as a gruesome way to kill an enemy, although scholars still disagree about whether the practice was ever actually done. Skaldic poems describe the ritualized killing in detail. King Aella was forced to lie face-down on a table. There, he was held in place while a slit was carved in his back to expose his ribs. His ribs were then cut open and his lungs pulled out so that they spread wide in the shape of an eagle's wings.

This method of killing, although referred to in several skaldic poems, may have been an exaggeration or poetic telling of the tale. Whether it was a literary invention of the skalds or an actual practice, there can be no doubt that, however he died, King Aella was forced to suffer.

Chapter 9

The Quest for Riches, Land, and Glory

When Bjorn Ironside came into the world, his father was already a rich king. Ragnar had been a Viking raider on a grand scale and had enough loot from Europe and England to last the sons for a lifetime. However, he had also begun the family custom of men going out into the world to find more riches and build a reputation for themselves.

The Norse people lived in a cold, inhospitable climate. Much of their land was above the Arctic Circle. There were animals to hunt and plants to gather, but the window of opportunity for acquiring them was small. When the winter winds blasted, the Vikings stayed close to home, eating the foods they had dried and set back during the warmer seasons. It was a dismal life, even for wealthy families like Ragnar's. The idea of going south to live must have been very appealing to the Vikings.

Because winters were so long, limiting wealth from farming, the Vikings were relatively poor. Bjorn Ironside, like his father, was one of the wealthiest and most advantageously positioned socially in the Viking realm. Yet, because the country as a whole was poor, there was little wealth to be gained in Scandinavia.

The Scandinavians had a skill that they could use to go out from their country and explore. Since they lived in a land split in many directions by water, they became master shipbuilders. These ships would take them farther than any other European ships had gone before.

So many Vikings left their homes that it stands to reason some of them may have gotten lost and never gained the riches and glory they sought or even made it back to their homes alive. Yet, enough of them did make it to other inhabited countries that they developed a reputation as world-class sailors and navigators.

They were searching for gold objects and coins they could make into ingots or bullion. They wanted the most prized material possessions from the countries they reached. Sometimes, they would set out for a specific location to get specific valuable objects. Other times, they would simply set sail and stop all along the way to take what they could find. Their raids were fierce and swift, terrifying the people of the countries they invaded.

Bjorn's father Ragnar had a close friend called Hastein. Hastein became Bjorn's mentor as he built his reputation as a naval commander. In about 860, Hastein and Bjorn sailed to the Mediterranean to plunder the lands along the Iberian coast, all the way to Gibraltar.

When Hastein and Bjorn Ironside reached the south of France, they pillaged it through the warm season and then spent their winter there, waiting for spring to arrive so they could continue with their journey.

Chapter 10

To Conquer Rome

When warm weather arrived, Bjorn Ironside and Hastein took their army and went inland. There, they captured the city of Pisa. While they were in Pisa, basking in the glory of their victory, Bjorn Ironside heard that the great city of Rome was nearby. Fascinated with the possibility of laying his hands on magnificent gold objects that had been built up there, he was ready to go and pillage Rome. And, just as important, raiding Rome would give him a reputation grander than even that of his father. And so it was that Bjorn Ironside's men left Pisa and hurried in the direction where they were told they would find Rome.

The city of Rome symbolized many things for the Viking people. First and foremost, it was a city filled with gold and other precious metals set aside for centuries. Any Viking raider who could take possession of all that wealth would have all he needed to provide a spectacular existence for himself and his family. He could melt down the precious metals to make ingots or bullion, which he could then trade for more land as well as all the things he and his family needed or desired.

Besides being a wealthy city, Rome was the seat of the Christian religion at that time. The Christians of the Viking Era were not kind to the Vikings. They disparaged the Vikings' pagan religion, saying it was devil worship. They spoke and wrote of the Norse people, saying they were ignorant, unclean and uncivilized. When the Christians took over in a Viking land, they outlawed the Vikings' customs and tore down or burnt the buildings where they held their rituals.

So, Rome was more than a place where riches could be gained. It was the home of enemies that sought to end their Viking civilization and replace it with their own religious customs and rituals.

Some Vikings were not opposed to the Christian religion or, as they called it, the new customs. In fact, as the years went by, more and more of the Norse people converted to Christianity.

However, Bjorn was not a Christian and had no intention of becoming one. To him, the people of Rome were at times an irritation and at other times a threat to his way of life. He wanted the riches, the land, and the glory, of course, but he also wanted the satisfaction of putting Rome in its place. The Vikings were ready to do some severe damage and gain vast wealth in the process. Excited for the battle to begin, Bjorn Ironside and Hastein gathered their men and headed further inland.

Chapter 11

A Plan Fulfilled?

Bjorn Ironside and his men reached the location where they had been told they would find Rome. They carried with them their heavy battle axes, spears, and swords. They not only had the battle equipment they needed, but they also had battle skills, honed since they were little Viking boys. They had wooden shields, some with iron edges, to defend themselves against anyone who dared to fight back. And, they had a fiery thirst for conquest and glory. It seemed they were sure to succeed.

However, when they reached the city, they could not break through its defenses. They fought long and hard, but Bjorn eventually realized that it was time to step back and make a new plan. Since they could not successfully invade the city in their usual head-on style, Bjorn reasoned that they must do it differently somehow.

Bjorn Ironside planned to get the Romans to invite him into the city. It would not be easy or straightforward, of course. After all, who would invite in a Viking, well-known for killing and looting in every European city they entered? Yet, if Rome and its people were too strong to overcome, the only way for the Vikings to proceed was for the priest there to allow it.

Bjorn's idea was to pretend he was near death. This would give the bishop of Rome the impression that he was weak and vulnerable. To interest the priest in his cause, he further pretended that he had converted to Christianity as many of the Vikings had already done. Bjorn Ironside wanted to receive the last sacraments of the church. What priest could resist?

Now, Bjorn was faking mortal illness, but of course, he was very strong and capable. To disguise this fact, he had himself carried to the priest in a coffin. Inside the coffin, Bjorn wore chain mail and held a mighty sword and battle-ax, ready to spring on the people of Rome when they least expected it.

The plan worked beautifully. When Bjorn Ironside leaped from the coffin, no one could have been more surprised than the priest. The small Viking guard that had carried Bjorn to see the priest quickly made their way back to the city gates, broke them open with their battle-axes, and let in the rest of Bjorn Ironside's Viking army.

From there, it was short work to kill the priests and other leaders and gain control of the city. Bjorn and his men fulfilled their quest to plunder Rome except for one little detail. The city they had breached was not Rome! Instead, it was the city of Luna, a smaller city north of Rome.

Discouraged after their failure to conquer one of the wealthiest cities in Europe, Bjorn Ironside, Hastein, and their army returned to Scandinavia. Bjorn Ironside had succeeded in proving his battle skills, bravery, and intelligent strategies. He succeeded in capturing the city he saw before him. If only he had been as good at finding his way on the land as he was at navigating the seas, he could have made an even greater name for himself – not only among his own people, but also on a global, historical scale.

Chapter 12

Björn Ironside, King of Sweden

When Bjorn Ironside and his brothers left England after killing Aella, they continued to fight and plunder. A list of the cities Bjorn Ironside raided would be long and include places like Spain, France, Italy Wales, England, Sicily, civilizations on the northern coasts of Africa, and of course, Luna.

As Bjorn and his fleet left Luna, they stopped at Sicily and North Africa for more plundering. However, when they reached the Straits of Gibraltar, they came upon the naval fleet of Al-Andalus. This enemy force sent the Vikings quickly on their way by bombarding them with Greek Fire, which was an incendiary weapon much like napalm. After 40 of his best ships were destroyed, Bjorn fled to Sweden and ruled as king there for the rest of his life.

How did Bjorn Ironside become King of Sweden? At the time of Ragnar's death, Lothbrok was the King of Denmark and Sweden, except that Ragnar had set up Eysteinn as king of upper Sweden. After Ragnar died, Bjorn and his brothers conquered Sweden, including the part of Sweden Eysteinn had ruled. The brothers divided up the land – Sweden, Uppsala and Denmark – and Bjorn became King of Uppsala and Sweden.

Uppsala was and is a part of Sweden. In Bjorn Ironside's time, Uppsala was the center of Viking pagan worship. Ruins have been discovered of a vast worship center where public rituals were held. The great hall was where rituals and their accompanying feasts were held. According to Adam of Bremen, an early Swedish king, the hall at Gamla Uppsala was lined with gold panels and had giant pagan statues throughout. In another building, next to the great hall, ritual sacrifices were held.

Artifacts from the pre-Christian era that were found at this large temple in Gamla Uppsala are in view today at the Gamla Uppsala Museum in Sweden. A new, Christian cathedral was later erected over the ruins of the pagan temple. However, in Bjorn's time, Sweden was primarily pagan. Because Uppsala and its temple were so important to Viking culture, Bjorn must have been very honored to have it in his realm.

Bjorn settled down in Sweden and lived the rest of his life as a wealthy and privileged king. The name of Bjorn's wife is not known, but he married her about 783 and had at least three children with her. He had two well-known sons, Erik, born in 798, and Refil, born in 796, both of whom were considered great warriors.

Chapter 13

A Viking Funeral

Some sources say Bjorn died in 859 in Uppsala, Sweden. Others give the same death date, but instead of Sweden, they give the location as Paris, France. Bjorn Ironside, wherever he died, would have been given a king's funeral.

Typically, many rituals accompanied the death of a Viking. First, ceremonies were held while the body was being prepared. One involved cutting the dead Viking's nails to ensure they were not used to build a ship that could go to Ragnarok, the final battle of the gods and the end of the human world.

One account of a Viking funeral, written by Ahmad ibn Fadlan, described the entire process from start to finish. First, the Viking was placed in a grave until his thralls could make new clothes for him. Then, he was placed in a longship. One of his women thralls voluntarily made his death journey with him. She was kept intoxicated for the entire ordeal.

When the time set for the cremation arrived, the other Vikings pulled in the longship. They placed the dead Viking on a large pallet of wood. An older woman who performed the death rite put cushions on the pallet, and the dead Viking was placed there in his new clothing.

Afterward, the slave girl had sex with each of the Viking men in each of their tents. Later, they attached her wrists and ankles to a door frame and held her up to see the afterworld, which she described in detail. They took her to the dead Viking, tied her up, and the old woman stabbed her.

The Vikings made several animal sacrifices to the gods and put the dead animals in the ship with the dead Viking. They also placed food in the ship to nourish the dead Viking on his way to the afterlife. They supplied him with mead or other alcoholic beverages as well.

Other accounts of Viking funerals tell how widows were sometimes sacrificed along with the slave girl. Other customs revolve around what must be done if anyone sees the ghost of the deceased Viking soon after his death. The Viking would have to die again. To accomplish that, a religious leader would put a stake through the corpse. If this was not done, the Vikings believed other family members would be put at risk and die.

When all this was completed, the Vikings took torches and set the ship on fire. When the ship had burnt to the ground, they built a barrow mound over it. Bjorn's barrow on the island of Munso is topped off with a runic device fitting for a king.

Chapter 14

The Munso Dynasty

Erik, known as Erik Bjornson, became the king of Sweden after Bjorn Ironside's death. Erik did not live as a king for very long. When he died, his brother's son, Erik Refilsson, took over the crown.

Often, two brothers would share the rule while they were both alive, so the record is not always clear. Furthermore, different Medieval texts about the region's kings take different perspectives. The *Gesta Danorum* by Saxo Grammaticus was weighted in favor of Denmark. The Ynglingatal is another account of the kings; it favors Sweden and the Swedish roots of the Norwegian kings. However, Bjorn's descendants certainly ruled Sweden for many generations as is recounted in all significant references.

By some accounts, Bjorn Ironside is considered a "Saga King." This epithet refers to the mythological and literary nature of the record of their kingdoms. Still, even those that include Bjorn as a literary invention show him as one of the last of the saga kings. As time goes by and more facts and artifacts are discovered, it becomes clearer and more apparent that Bjorn not only lived, but he was a King of Sweden and the founder of a dynasty of rulers in the Kingdom of Sweden.

Bjorn Ironside's mother Aslaug was the daughter of Sigurd, who was said to be a descendant of Odin. Therefore, all of the Munso Dynasty and all of its descendants are seen as coming from Odin's line. Odin was a Norse god associated with healing, battle, sorcery, knowledge, poetry and the Runic alphabet. This bit of lore about the House of Munso being descended from Odin increased the people's respect for the Munso Dynasty, as it was seen as being related to the mythological gods the people revered.

Odin was depicted in the Old Norse texts as a man with a long beard and a broad hat. He carried a staff and rode on a flying steed into the underworld. He continually sought to attain more knowledge and often appeared in disguise. Humorously, Odin was said to bet with his wife Frigg over the outcomes of the people's exploits.

His name is included in the Norse creation myth, in which he slays Ymir, a primordial being, and gives gifts to the first two humans. He is closely associated with Yule, a holiday still kept by many modern pagans. And, he was said to have many, many sons, one of whom was thought to be the ancestor of Aslaug and so, of Bjorn Ironside.

Centuries after King Bjorn Ironside died, a mound assumed to be Bjorn Ironside's burial place, was found on the island of Munso, leading his dynasty to be called the Munso Dynasty. While scholars sometimes include Sigurd Ring, Ragnar Lothbrok and Eysteinn Beli in the Munso Dynasty, most others count Bjorn Ironside as the first king of that dynasty. This may be because Bjorn Ironside was the first of these kings to have a firm foothold in the historical record. Virtually all historians agree, though, that Bjorn Ironside's descendent named Eric the Victorious, who died in 995 CE, has been well proven as an actual person who was indeed a king of Sweden. From that Eric on, the record is abundantly clear.

Other names for the Munso Dynasty include the House of Ivar Vidfamne, the House of Uppsala, and the Old Dynasty. When, after a time of civil war, the Munso Dynasty was expelled from

Sweden, it started back up in Denmark and became the ruling house there.

Chapter 15

Björn Ironside's Story Lives On

Bjorn Ironside lived in a time and place where stories were passed on in the oral tradition rather than through written texts. So, how is it possible that we can know anything about Bjorn Ironside's life after more than a millennium has passed? The first answer is that the stories were composed and memorized by the scalds and told in the same way, time and time again.

In Medieval times, those stories were written down. Several of these texts have survived, been translated, and continue to be studied by scholars even today. Below is a partial list of Medieval texts that contain elements of Bjorn Ironside's story.

Hervarar Saga

This story comes from the 1400s, but it contains much earlier material from the Skaldic tradition of the time of Bjorn Ironside. The story follows what happens with the sword Tyrfing – how it was used in battles, ended up in a barrow or burial mound, and was retrieved by the shield-maiden Hervor. Near the end of the Saga are the stories of the Swedish kings, including Bjorn Ironside.

Saga of Ragnar Lothbrok

The Saga of Ragnar Lothbrok is an Icelandic text that was composed in the 1400s and based on the Skaldic stories and poems. In addition to telling the story of Ragnar and his wives, the story goes into accounts of Ragnar's death and his sons' reaction to it. It follows many of the battles in which Ragnar's sons, including Bjorn Ironside, fought.

Tale of Ragnar's Sons

After the Tale of Ragnar's Sons deals with the life of Ragnar Lothbrok, it goes on to follow the stories of Ragnar's children. It tells how they exacted their vengeance on King Aella and how they led the Great Heathen Army in battles around Europe. Finally, it tells the stories of the Scandinavian Kings, one of which was Bjorn Ironside.

Heimskringla Saga

Snorre Sturlsun wrote this saga about the Old Norse kings in Iceland. It tells the stories of Viking Kings.

Icelandic Sagas

The Heimskringla Saga was one of a larger body of stories known as the Icelandic Sagas. These were written down by various chroniclers. One of them may have been Snorri Sturlsun. These are the stories that were told in Iceland immediately following their arrival there. They told the tales of the important Vikings who went before them and lived in Scandinavia. They are full of drama, blood, sex, and the exploits of the Viking raiders. They form the basis of many of the tales told today about the Vikings of the time of Ragnar, Bjorn and his brothers, and their sons. The people of Iceland were and are great storytellers. In fact, today 1 in 10 Icelanders is a published author.

The descendants of the Vikings are, of course, interested in learning about their ancestors. Most of the scholars studying these subjects live in Scandinavia. However, as interest in the Vikings has grown and filtered down to the masses, more and more fictional accounts of the Viking Age are being written.

Modern depictions of Viking times have been presented in movies, television series, and books. The time of Bjorn Ironside was a period when many of the customs and lore seem strange to modern viewers and readers. Thus, the tale has fascinated audiences and delighted readers since these fictional accounts began to be written.

J.R.R. Tolkien used many of the names, locations and cultural practices of the Scandinavian people for his creation of Middle Earth. These stories originated with the Skalds before they were made a part of the written historical reference.

Currently, interest in the Vikings of Bjorn Ironside's day has increased dramatically. After all, more is known now than in the early days of movies and television. As more archeological digs uncover more evidence of the practices, customs, and history of the Norse people, scholars are eager to learn even more.

Always remember, though, that fictional accounts often change the known facts to make their scripts or manuscripts more dramatic and more relatable to modern audiences. As long as you consider the source, you can still find great pleasure in these fictionalized tales. For accuracy, always go back to the Medieval texts, the scholars' translations and interpretations, and the archeological record. But, when it comes to being engrossed in this compelling story, you may find current movies and books more fun and enjoyable.

And, as the interest among scholars grows, the fascination is dripping down into the general population. Even more popular fiction is sure to be written as the genre blossoms and grows.

Conclusion

There can be no doubt about the power of the story of Bjorn Ironside. It inspires us to follow the examples of our parents and move beyond them to find greater success than our parents ever imagined. It encourages us to take the initiative to move beyond our physical homes and into the wider world.

But, there are also caveats within the Bjorn Ironside story. There are failures to understand and learn from throughout the texts that describe Bjorn's time and still apply to us today. The result of Ragnar's jealousy towards his sons is one example we can study to understand the destructive nature of familial envy.

Reading or watching Bjorn's story brings knowledge of a different culture and a different time that is so far removed from our customs that we must stretch our minds to follow and understand it.

After getting a brief introduction to Bjorn's story, you can continue to learn more by reading translations of the old texts, which are compelling in themselves. While many students find history a dreary parade of names, locations, and dates, the historical accounts of Bjorn Ironside and his brothers go well beyond the simple facts to reveal the more profound significance of the stories and the motivations behind the characters' actions.

If you ever have a chance to speak to a scholar studying the Old Norse myths and legends, you can add to your store of knowledge on Bjorn Ironside's life and times in a simple conversation. Perhaps you will even be inspired to take up the study yourself.

Keep watching for new additions to the story as archeological evidence accumulates and fictional accounts explore the world of the Vikings in greater detail.

Bjorn Ironside was an extraordinary person, a mighty Viking warrior, an early King of Sweden, and the start of the Munso Dynasty. Some scholars refer to him as the Viking Ideal because of his success in battle, and because of his ultimate decision to settle down and rule generously. Whatever he was, the story of Bjorn Ironside is one worth reading, studying, and passing on to others. Was Bjorn Ironside a real person? If not in a factual way, at least in terms of meaning and heroic legend, he was more real than most mortals who have walked the earth before or since.

Made in United States
Troutdale, OR
09/17/2024

22919804R00040